Eyes on the Sky

Mercury

by P.M. Boekhoff and Stuart A. Kallen

KIDHAVEN
PRESS™

THOMSON
✶
GALE

San Diego • Detroit • New York • San Francisco • Cleveland
New Haven, Conn. • Waterville Maine • London • Munich

For more information, contact
KidHaven Press
27500 Drake Rd.
Farmington Hills, MI 48331-3535
Or you can visit our Internet site at http://www.gale.com

LIBRARY OF CONGRESS CATALOGING-IN-PUBLICATION DATA

Boekhoff, P.M. (Patti Marlene), 1957–
  Mercury / by P.M. Boekhoff and Stuart A. Kallen.
    p. cm.—(Eyes on the sky)
  Includes bibliographical references and index.
  Summary: Discusses the mysteries of the planet Mercury, its relationship with the sun and other planets, its geography and climate, and the physical properties of the planet.
  ISBN 0-7377-1000-4 (hardback : alk. paper)
  1. Mercury (Planet)—Juvenile literature. [1. Mercury (Planet)] I. Kallen, Stuart A., 1955– II. Title. III. Series.
  QB611 .B64   2003
  523.41—dc21

2001005693

Printed in the United States of America

Table of Contents

1
Mysterious Mercury

Mercury is one of seven planets that can be seen from Earth without the aid of a telescope, but Mercury is not easy to find. It never appears high in the night sky, where planets are easiest to see from Earth. Mercury appears near the sun, and it is so close that it almost always disappears into the sun's bright glare.

Because Mercury is so close to the sun, it can be seen from Earth only about thirty-five days each year. It is most visible when it appears beside the rising or setting sun. At those times the planet looks like a small, bright yellow-orange dot appearing on the **horizon**.

Although Mercury was difficult to see, ancient **astronomers** kept track of its activity for thousands of years. They explained the movement of the planets as the actions of gods that lived in the sky. More than five thousand years ago, the ancient Egyptians called Mercury Sobkou, and they discovered that the planet circles around the sun. Ancient Babylonians called it Nabou, meaning the ruler of the universe. Nabou was said to wake the sun every morning and move it across the sky. The ancient northern Europeans called Mercury Odin or Wodin. As Wodin, Mercury was the wise father of the gods. He was said to be a traveler who followed the ancient trade routes, bringing messages, knowledge, and goods to and from distant cultures.

Mercury, a small yellowish-white dot, appears to the left of the new crescent moon.

Ancient Greeks named the planet after Hermes, the messenger of the gods.

Because Mercury traveled faster than the other planets, the ancient Greeks named it Hermes, the messenger of the gods, because it appeared to fly very fast across the sky. Hermes was the god of twilight, the soft light from the sky during early morning or early evening when the sun is hidden but its light remains. This is the time when Mercury may be seen from Earth. As messenger of the gods, Hermes announced the rising of Zeus, the god of day.

Ancient astronomers knew that Mercury traveled quickly because it returned to the same position in the sky in less time than other planets. But later some of the Greek astronomers became confused when they saw Mercury on one side of the sun in the morning, and on the other side in the evening. Because of this, some people thought Mercury was two different stars. When Mercury appeared in the morning they called it Apollo, and in the evening, Hermes. The Romans later adopted the Greek god Hermes, renaming him Mercurius, now called Mercury.

Aristotle's Universe

Unlike stars, which twinkle, planets glow with a steady light. In about 350 B.C., Greek astronomer Heracleides of Pontus observed that the two stars were actually one planet traveling around the sun. In the next century, Aristarchus of Samos added that Earth and all the other planets also circled around the sun. But most astronomers did not accept the knowledge

Plato (left) and Aristotle discuss the theory of an Earth-centered universe.

of Aristarchus and others. They agreed with the philosopher and scientist Aristotle, who believed that Earth was the center of the universe.

About eighteen hundred years later, between 1530 and 1543, Polish astronomer Copernicus rediscovered the ancient knowledge that Earth, Mercury, and all the other planets circled around the sun. Although he never saw Mercury through the thick mists near his home in Poland, Copernicus used other people's

Polish astronomer Copernicus reaffirms that all the planets revolve around the sun.

records of observations to determine that Mercury takes eighty-eight days to **orbit** the sun.

Because the time it takes a planet to travel once around the sun is equal to one year, on Mercury one year is equal to eighty-eight Earth days. This the shortest year of any planet in the solar system.

Through the Telescope

In the 1600s astronomers used the first telescopes to learn more about Mercury. Men such as Italian astronomer Giovanni Zupus kept track of Mercury's movements and used mathematics to study the planet's orbit. He and others established that Mercury is the innermost planet in the solar system, and that it has a very odd orbit.

In 1639 Zupus looked at Mercury through a powerful new telescope. He saw that part of the planet is in shadow, and part is lit by the sun. Zupus discovered that the lighted part of Mercury changes from a crescent sliver to a full circle. These changes, which repeat over time, are called **phases**.

Astronomers began to search for other changes, or phases, in mysterious Mercury. They wondered whether the planet always kept the same side facing the sun, or whether it spun around in circles on its own **axis**. To find out whether Mercury twirled around as it orbited the sun, astronomers tried to find a marking, such as a huge mountain or crater, on the surface of the planet. If they could find such a marking, they could determine how long it took before that marking appeared again. Then they would know how long it took Mercury to spin around one time.

Mercury could not be seen very well through a normal telescope and for centuries, astrono-

The sun lights a crescent sliver of Mercury.

Powerful telescopes like the Very Large Array radio telescope in New Mexico allow astronomers to see Mercury's rotation.

mers could not find a good mark to use to time Mercury's **rotation**. It was not until 1965 that astronomers were able to use a high-powered radio telescope, called a radar telescope, to bounce radio waves off the surface of Mercury. The images sent back to Earth by the radar telescope showed the markings on Mercury. Astronomers finally proved what many had suspected for years. Mercury slowly spins around as it orbits the sun.

Mercury has fascinated and confused astronomers ever since ancient times. Because it is hidden in the glare of the sun, it took more than five thousand years to accurately track the planet's orbit and rotation. Even with the high-tech tools of the modern space program, there is much that remains unknown about the planet. Like so many before them, astronomers will continue to study the planet, searching for answers to the mysteries of Mercury.

2
Mercury and the Sun

Earth rotates on its axis every twenty-four hours, making one whole Earth day. But Mercury rotates very slowly on its axis, turning only once every 58.7 days. Earth orbits the sun in about 365 days to make one Earth year, but Mercury takes only 88 days to orbit the sun. When the motions of Mercury's rotation and orbit are combined, a very odd event happens.

One day on Mercury, from sunrise one day to sunrise the next day, stretches out to almost 176 Earth days. But Mercury's year, the time it takes to orbit the sun, is only 88 Earth days long. This means that on Mercury, a day is twice as long as a year!

Mercury turns slowly on its axis at a steady speed, but its quick orbit around the sun is not steady. Mercury's orbit is **elliptical**, meaning

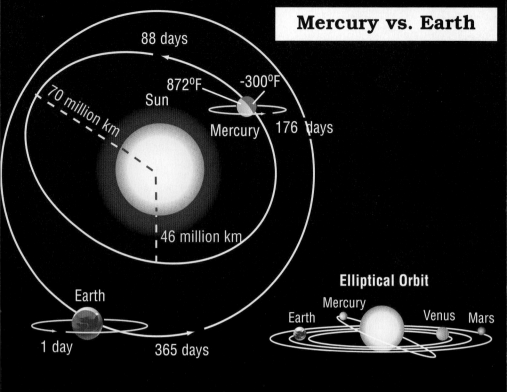

88 days

872°F

-300°F

Sun

Mercury

176 days

70 million km

46 million km

Elliptical Orbit

Earth

Mercury

Earth

Venus

Mars

1 day

365 days

that it is oval-shaped rather than round. When Mercury comes closer to the sun in its elliptical orbit, it moves faster.

Mercury also orbits at an odd angle to the sun. Most of the planets in the solar system line up in a plane, or flat disk shape, when traveling around the sun. Mercury dips above and below the plane of planetary orbits more than any other planet except Pluto, the outer-most planet.

Because Mercury is not always the same distance from the sun, and its speed and angle of orbit is changing, the sun appears to move across Mercury's sky in an uneven pattern.

From Mercury, sometimes the sun will seem to rise, then set for a short time, then rise again. The same thing sometimes happens at sunset—the sun appears to set, rise again for a short time, then set again.

Einstein Figures it Out

For many years, astronomers wondered why the orbit of Mercury is so unusual. They believed it had something to do with **gravity**, which is the force of a star or planet that draws objects toward its center. Each planet, and stars such as the sun, pulls with its own force of gravity. Astronomers figured that Mercury is held in its orbit by the gravity of the sun. The other planets in the solar system also pull on Mercury with their force of gravity, causing it to have an odd orbit around the sun.

But when astronomers added all the forces of gravity pulling on Mercury, there was still some unusual motion to the planet they could not explain. The point where Mercury passes closest to the sun changes a little bit every time it orbits. Mercury's motion was finally explained by physicist Albert Einstein, when he worked out a new theory of gravity called the theory of relativity. Einstein proved that because it is so close to the sun, Mercury experiences a different kind of gravity than every other planet in the solar system.

Albert Einstein proved that Mercury experiences a different kind of gravity than the other planets because of its closeness to the sun.

Weak and Strong Gravity

Before Einstein, only one kind of gravity was known—the kind experienced on Earth. Einstein called this weak gravity. Weak gravity keeps everything on the surface of the planet from floating away into outer space. The larger the planet or heavenly body, the more gravity it has. Mercury has less gravity than Earth because it is much smaller, less than half the size of Earth. On larger planets with more gravity, objects have more weight. For example, a girl who weighs 80 pounds on Earth would weigh less than half that, about 30 pounds, on Mercury. She could jump high into the air because there would be less gravity pulling her down.

The sun is many times larger than all the planets in the solar system put together. Because of this, it has the strongest gravity. Einstein proved that the pull of gravity close to the sun is so strong that it pulls down even the space around it. Space is drawn down into a curved shape around the sun. Mercury travels in the space that is curved by the strong gravity of the sun. Mercury's odd orbit and fast motion is caused by its closeness to the strong gravity of the sun. If Mercury did not speed up when it was closest to the sun, the sun's gravity would drag Mercury into its massive blazing inferno

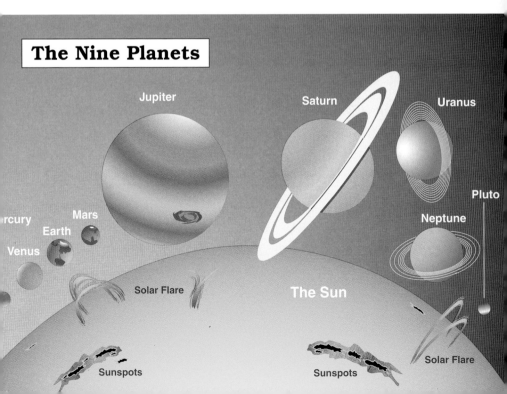

The Nine Planets

Jupiter

Saturn

Uranus

Mercury

Mars

Earth

Venus

Neptune

Pluto

Solar Flare

The Sun

Sunspots

Sunspots

Solar Flare

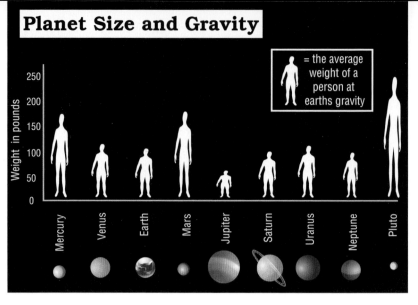

Planet Size and Gravity

and burn it up. Mercury's speed and odd orbit help it to resist the strong gravity of the sun.

Atmosphere

Large planets have enough gravity to hold on to gases that surround them and make up their **atmospheres**. For example, Earth's atmosphere, which is held in place by gravity, is made up of oxygen, nitrogen, and other gases. Mercury's atmosphere contains tiny bits of hydrogen, helium, oxygen, sodium, and potassium which seep out of its **crust** or are blown across it by the sun. Mercury, however, does not have enough gravity to hold on to these gases, so most of its atmosphere escapes into space. Earth's thick life-giving atmosphere contains the air people breathe, but Mercury's atmosphere is much too thin to support human life.

Earth's atmosphere also scatters the light from the sun, lighting up the entire sky in the daytime. Because there is not enough air to scatter the light on Mercury, the sky is always black. In the daytime, the sun looks two and a half times as big as it does on Earth. A small halo of yellow light glows around the blazing sun, surrounded by distant stars in the black sky. Two bright planets in the Mercury sky glow with a steady light: the cream-colored planet Venus and the blue-green planet Earth.

Hot and Cold

Earth's thick atmosphere protects it from extreme heat, X rays, ultraviolet light, and other forms of solar energy that could kill human beings. At night, Earth's atmosphere holds in some of the heat that is collected from the sun during the daytime, keeping the temperatures from becoming too cold. Mercury's atmosphere is too thin to protect it from the extreme heat and radiation of the sun in the daytime, or to hold the heat near the planet's surface at night.

With little atmosphere, the surface temperature on the side of Mercury that faces the sun—the day side—reaches up to about 872° Fahrenheit (F). This is twice as hot as a kitchen oven, and hot enough to melt soft metals. On the dark side of Mercury facing away from the sun—the night side—the temperature dips to about minus 300° F. This temperature is much colder than the North Pole.

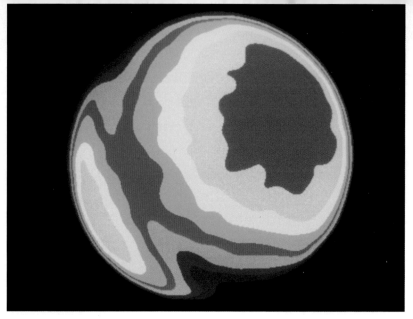

A computer-enhanced photo shows the day side of Mercury in red where temperatures reach 872º F.

The coldest temperature on Earth is about minus 100° F, and the hottest is about 130° F, a difference of about 230° F. Mercury's temperature changes more than one thousand degrees between night and day, the most extreme difference on any planet. While half of the planet bakes in the hot sun, the half facing away from the sun freezes in a long, cold night.

With its closeness to the sun, thin gravity, and lack of atmosphere, Mercury will never be fit for human life. But scientists continue to study the planet hoping to find out more about the sun's gravity, heat, and other aspects. By learning more about this planet with a wobbly orbit, weak gravity, and extreme temperatures, astronomers may find out more about the sun, Earth, and the stars beyond.

3
Fire and Ice

For at least five thousand years, the surface of Mercury was a complete mystery to astronomers. This changed on November 3, 1973, when the National Aeronautics and Space Administration (NASA) launched *Mariner 10,* an unmanned space probe with a camera onboard. On March 29, 1974, *Mariner 10* passed within 170 miles of Mercury's surface. *Mariner 10* then went into orbit around the sun, passing Mercury two more times at a distance of about 200 miles.

Mariner 10 photographed about 45 percent of Mercury's surface, which was lit by the sun during the flybys. The pictures taken by *Mariner 10* reveal everything that is known today about the surface of Mercury. By studying the photographs, astronomers can estimate

A computer image depicts the *Mariner 10* sending photographs of the surface of Mercury back to Earth.

how the surface of Mercury may have been formed in the distant past.

Ancient Battered World

Billions of years ago giant chunks of stone and iron called **meteoroids** tracked through the ancient skies. Other objects, called **comets**, made of water, gases, and dust, hurtled around the newly formed planets. Some of these space rocks crashed into Mercury, forming large round holes on the surface, called craters. When this happened, rocks were thrown up and part of the planet's surface (and possibly part of its atmosphere) was stripped away, leaving a very thin crust of rocks and dust.

Earth was formed in much the same way, but it formed a thick atmosphere to protect it and hold large amounts of water to its surface. The atmosphere of Earth burns up most comets and meteoroids before they hit the planet. The water and weather in the atmosphere of Earth wear down the surface of ancient craters until they are almost gone. Because Mercury has almost no atmosphere, it has no weather to protect or change the surface.

Craters

Pictures sent back to Earth by *Mariner 10* show that Mercury's surface is very similar to the surface of Earth's moon—a small world, where the atmosphere is very thin and craters still mark the places where worlds once collided. In fact, Mercury looks like a copper-colored version of the moon. Both have millions of craters, some of them surrounded by bright rays that spread out from their centers. These are caused by debris that flew out from the craters when rocks called

Artwork portrays the heavily cratered, copper-colored surface of Mercury. The inset photo shows actual craters photographed by *Mariner 10*.

meteorites fell on the planet from outer space. The rayed craters were made millions of years ago by the last large meteorites to hit Mercury.

Because Mercury is made of harder, heavier materials than the moon, its craters were probably never as deep as the moon's craters. The surface was probably smoothed out even more by melting that occurred when hot **asteroids** (or small planets) and meteors crashed and burned on the planet. Later, hot lava from volcanoes may also have played a part in smoothing the dusty hillsides along the edges of the older craters.

Caloris Basin

The smallest craters that *Mariner 10* was able to photograph on Mercury are a few hundred feet wide. The largest known crater on Mercury, and one of the largest in the solar system, is called the Caloris Basin. Caloris is truly gigantic. It is more than eight hundred miles wide and is surrounded with rings of mountains up to a mile high.

Caloris, which means "heat" in Latin, is also the hottest spot on the planet, sometimes reaching more than 800° F. Caloris faces the sun when Mercury's orbit brings it closest to the sun. On the opposite side of the planet from Caloris is another hot spot. In this area, the surface seems to be covered with patterns of shock waves, perhaps created from the same impact that created Caloris.

Astronomers think Caloris was caused by a huge asteroid that crashed into Mercury 4 billion years ago. When an asteroid crashes into a planet, the crust around the crash site moves out in rippling waves.

Scarps

The ripples caused by asteroids and meteorites are called scarps. Caloris is so large that it must have been caused by a very big crash, sending shock waves through the entire planet and causing scarps to form on the other side.

Scarps are long lines of cliffs that crisscross the surface of Mercury. From above, the scarps look like big cracks and wrinkles. Astronomers believe that these scarps began to form long ago when the interior of the planet slowly cooled and shrank. Its thin, rocky crust became too big for its shrinking center, so it wrinkled and cracked like the skin on an old dried-up potato. When asteroids and other space rocks hit the wrinkled crust, huge blocks of Mercury's crust were thrust upward along the cracks, forming giant cliffs along the fault lines.

On one side of the scarp the land is raised, and on the other side the land is lowered. Long, winding scarps pass across mountains, craters, and plains alike, splitting them into higher and

An image from the *Mariner 10* mission shows the Caloris Basin, Mercury's hottest spot.

lower parts. Some scarps are one to two miles high, and some are hundreds of miles long. A scarp called Discovery is one of the largest, at almost two miles high and more than three hundred miles long.

Volcanic Plains

In contrast to the jagged scarps, Mercury's surface also has large, gently rolling hills and flat, dusty plains. Some of the dust that covers

Mercury's hills and plains crumbled out of rocks cracked open by extreme temperature changes. The hills and plains themselves are shaped like the lava that flows from volcanoes.

Lava smoothes out the crust by flowing into the low places on the surface. Smooth plains are found in many areas of Mercury, such as around the outside of Caloris Basin. Some of the craters on Mercury are smooth inside as well. Inside the Caloris Basin, the surface has formations that look like curvy, winding river valleys of flowing lava, cooled and hardened many years ago. Arrow-shaped forms, typical of lava flows, also appear on Mercury.

Mercury's Geography

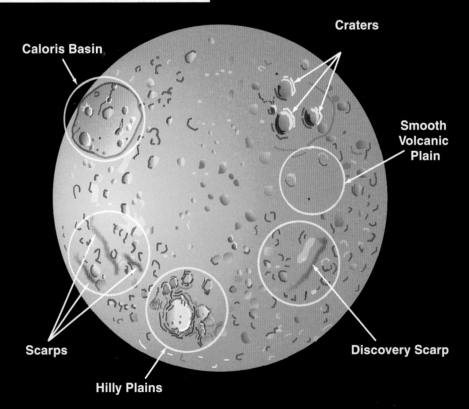

Craters

Caloris Basin

Smooth Volcanic Plain

Scarps

Discovery Scarp

Hilly Plains

Similar to Mercury, lava flows from a volcano, smoothing out Earth's surface.

Though no one has ever seen a volcano on Mercury, lava may have spewed out from the planet's hot **core** (center) billions of years ago. The flow of hot lava may have formed large, smooth plains on the other inner planets— Venus, Mars, and Earth—and on the moon as well. Large, active volcanoes still exist on all except the moon.

Scientists still do not know if Mercury was partially formed by lava flows. They also do not know what the planet and its atmosphere are made of. *Mariner 10*'s photographs captured less than half of Mercury and were taken from about two hundred miles away. Although the space explorer has revealed much about Mercury, there is still much to learn about the rocky, dusty surface of the closest planet to the sun.

4
Magnetic Mercury

A stronomers do not know for sure how Mercury and the solar system were formed, but they have a strong theory. About 5 billion years ago, stars exploded, leaving a giant cloud of dust and gas. The dust and gas swirled around, shrinking into a slowly rotating sphere. As the sphere became heavier, it collapsed under its own weight and flattened into a disk. As it flattened, it spun faster and faster until a star, our sun, ignited in the center.

The lighter space materials were boiled away, moving far from the sun. Heavy metal dust gathered close to the sun, colliding and sticking together as it swirled in a ring around the burning star. The heaviest metals, such as iron, orbited closest to the sun, forming Mercury's massive metallic core.

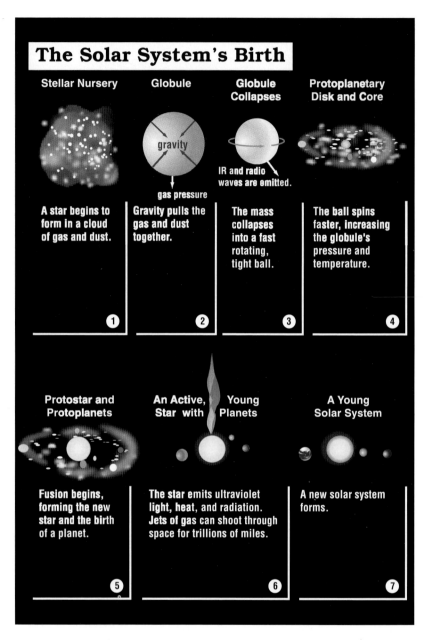

The Solar System's Birth

Stellar Nursery	Globule	Globule Collapses	Protoplanetary Disk and Core
A star begins to form in a cloud of gas and dust.	Gravity pulls the gas and dust together.	The mass collapses into a fast rotating, tight ball.	The ball spins faster, increasing the globule's pressure and temperature.
1	**2**	**3**	**4**

gravity

gas pressure

IR and radio waves are emitted.

Protostar and Protoplanets	An Active, Star with Young Planets	A Young Solar System
Fusion begins, forming the new star and the birth of a planet.	The star emits ultraviolet light, heat, and radiation. Jets of gas can shoot through space for trillions of miles.	A new solar system forms.
5	**6**	**7**

Of all the planets in the solar system, Mercury seems to have the largest core, about 65 percent of the planet. Around the giant metal core is the **mantle**, a layer of liquid rock similar to lava. A thin skin of solid rock floats atop this layer, forming the surface, or crust, of the planet.

Astronomers believe that Mercury's rocky crust holds heat below the surface in the core of the planet. The crust does not allow the heat to escape from the core in spite of the very cold temperatures on the dark side of the planet.

Magnetic Field

Astronomers believe that the core of Mercury is largely made of iron, like the core of Earth. When Mercury and Earth were younger and hotter, the metal in their cores was liquid. As the planets cooled, the center of their cores began to harden and shrink. On Earth, the outer part of the core stayed hot and liquid. Astronomers think that Mercury's outer core may also still contain some hot liquid metal.

As Earth rotates, it causes the liquid iron in the outer core to swirl around, creating a global **magnetic field**. Mercury is thought to have a global magnetic field about 1 percent as strong as the magnetic field surrounding Earth. This means that Earth and Mercury both behave like giant magnets, though Mercury's magnet is much weaker than Earth's.

Mercury is the only **terrestrial** planet besides Earth that is thought to have a global magnetic field. It is possible that Mercury's magnetic field is weaker because it rotates

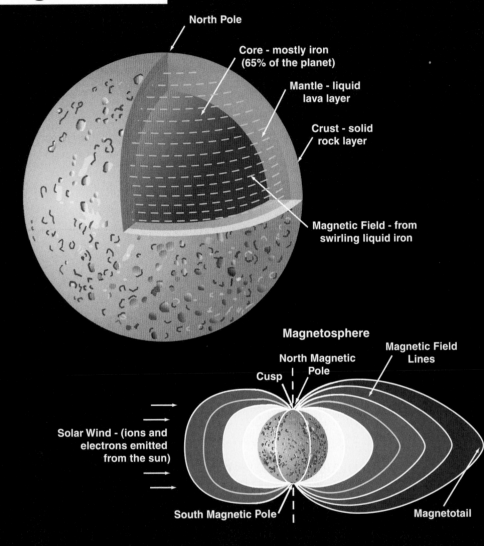

North Pole

Core - mostly iron
(65% of the planet)

Mantle - liquid
lava layer

Crust - solid
rock layer

Magnetic Field - from
swirling liquid iron

Magnetosphere

North Magnetic
Pole

Magnetic Field
Lines

Cusp

Solar Wind - (ions and
electrons emitted
from the sun)

South Magnetic Pole

Magnetotail

more slowly than Earth, so the hot metal in its outer core does not swirl around as fast. Or maybe it is more solid than Earth's core, and it has less liquid metal to swirl.

On both planets the **poles** of the magnet lie near the north and south poles. This means that on Mercury, as on Earth, a compass needle will point north. Because only 45 percent of Mercury has been mapped, however, astronomers are not sure whether Mercury's magnetic field surrounds the whole planet like the magnetic field around Earth.

If Mercury's core has cooled enough to become completely solid, it might have magnetic rocks left over from an old, dead magnetic field. These rocks would create local magnetic fields, such as those found on the moon and Mars. Local magnetic fields may be too weak or small to connect with other local magnetic fields. Because of this, the magnetic fields of Mercury may not surround the entire planet.

Polar Ice Caps

On the north and south poles of Mercury, the sun barely rises above the horizon. When the poles of Mercury are looked at through a radar telescope on Earth, some areas reflect like ice. They are thought to be areas inside deep craters where the sun never shines. Such areas would always be in shadow, too dark for cameras such as those on *Mariner 10* to see.

The shiny spots on Mercury are not large like the ice caps at the north and south poles of Earth. They are spotty and small, and there are

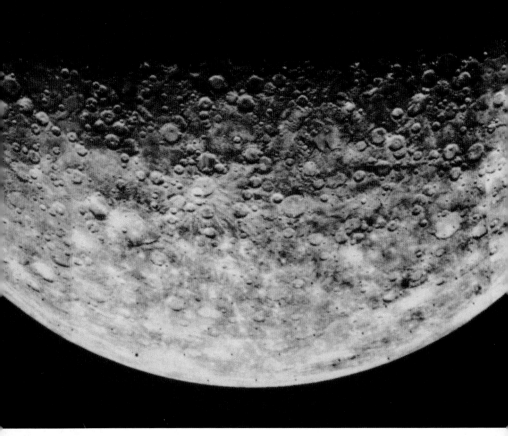

Dozens of shiny craters on the south pole of Mercury reflect light like ice.

dozens of them. The largest shiny spots on Mercury are around crater-sized areas near the poles. Crescent-shaped shiny spots appear in areas where the shadows of tall cliffs seem to block out the sun.

Astronomers think that the temperature in the shady zones of Mercury would be colder than minus 260° F. This temperature would be cold enough to preserve a patch of ice for billions of years. If the shiny spots are ice, it may have been left on Mercury's surface by a comet.

Ice and freezing clouds circling the poles of Mars are shinier than the bright spots on Mercury.

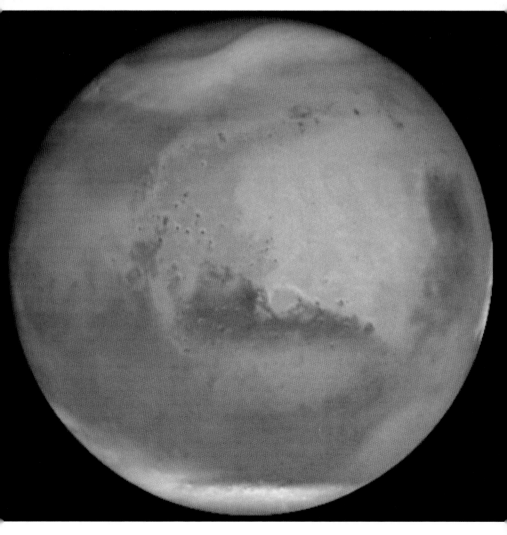

The crash of one big comet would have been enough to create the small amount of ice that appears to exist on Mercury. Water may also have been pushed from inside the planet long ago, then trapped in the freezing shady zones.

But the bright spots may not be iced water. Radar images show that the spots are not as shiny as ice on other planets. It is shinier, however, than the silicate rock that makes up most of Mercury's surface. It might also be shiny metal dust, or very thin patches of ice on the surface dust, or ice covered with a thin layer of dust. Many questions remain to be answered about these mysterious shiny patches.

MESSENGER

MESSENGER is the name of the next NASA flight to Mercury, scheduled for launch in 2004. The letters in *MESSENGER* stand for MErcury Surface, Space ENvironment, GEochemistry and Ranging. *MESSENGER*'s mission is to make detailed maps of the entire planet and measure the atmosphere and the magnetic field.

Tiny instruments onboard *MESSENGER* will try to answer questions about the mysterious planet. *MESSENGER* will also determine the planet's density, composition, and structure of its crust, and whether the planet was shaped by volcanoes. Astronomers will also try to analyze Mercury's atmosphere, the strength

An artist's view of *MESSENGER* as it nears Mercury.
Scientists hope *MESSENGER* will answer their
questions about the sun's nearest neighbor.

and makeup of its magnetic field, and what the shiny substance is near the poles. They will try to prove or disprove theories, such as the theory that Mercury, after being formed out of the sun, became a moon of Venus, then was knocked out of its orbit around Venus and into its present orbit.

Mercury has helped scientists form important ideas about the solar system. Its movements helped Copernicus understand that the planets circle the sun and helped Einstein explain his theory of relativity. Many astronomers believe that the planet holds answers to questions still unanswered about our solar system. Understanding Mercury and the forces that have shaped it will especially help astronomers to understand all the terrestrial planets, including Earth.

Glossary

asteroids: Small planetlike objects that revolve around the sun.

astronomer: A scientist who studies objects in space.

atmosphere: The gases around a heavenly body.

axis: An imaginary line that a planet spins around.

comet: Icy space rock with gaseous tails.

core: The center of a planet.

crust: The surface of a planet.

elliptical: Shaped like an oval, or an egg.

gravity: The force that pulls things toward the center of a planet, moon, or star.

horizon: Where Earth and sky appear to meet.

magnetic field: The area around a magnet or an electric current.

mantle: The layer between the crust and core of a planet.

meteorites: Metal or rocks that fall to a planet from outer space.

meteoroid: A solid body moving through space that is smaller than an asteroid and at least as large as a speck of dust.

orbit: The path of travel around an object.

phase: The repeated change in appearance of a planet or moon.

pole: Either end of an axis through a sphere (ball shape).

rotation: Spinning on an axis.

terrestrial: Like Earth and its land.

For Further Exploration

Books

Isaac Asimov, *Mercury: The Quick Planet*. Milwaukee: Gareth Stevens Publishing, 1989. Isaac Asimov describes quirky Mercury.

Duncan Brewer, *Mercury and the Sun*. New York: Marshall Cavendish, 1992. This book describes Mercury in relation to the sun and other planets.

Larry Dane Brimmer, *Mercury*. Chicago: Childrens Press, 1998. Describes Mercury and the pictures taken by *Mariner 10*.

Robert Daily, *Mercury*. New York: Franklin Watts, 1994. Describes Mercury and tells the story of its discovery.

Amy Margaret, *The Library of Planets: Mercury.*

New York: PowerKids Press, 2001. Describes Mercury's history, exploration, and features.

Seymour Simon, *Mercury*. New York: Morrow Junior Books, 1992. Describes what the *Mariner 10* photographs show about Mercury.

Websites

MESSENGER Home, (http://sd-www.jhuapl. edu). A site with many great links run by the scientists working on the *MESSENGER* space mission, scheduled to visit Mercury in 2004.

NASA Just for Kids, (www.nasa.gov). A site run by the U.S. space agency with information, photos, videos, art, and links relating to the solar system.

Index

Picture Credits

Cover and Title Page photo: © Roger Harris/Science Source/Photo Researchers (main); Hulton/Archive by Getty Images (inset)

© Paul Almasy/CORBIS, 8

© CORBIS, 24 (inset)

© Corel Corporation, 29

Hulton/Archive by Getty Images, 22

© David A. Hardy/Science Source/Photo Researchers, 24 (main)

© Roger Harris/Science Source/Photo Researchers, 10

© The De Morgan Foundation/Bridgeman Art Library, 6

Chris Jouan and Martha Schierholz, 17

© Dr. Michael J. Ledlow/Science Source/Photo Researchers, 20

© Dr. David Millar/Science Source/Photo Researchers, 5

NASA, 11, 39

© NASA/Roger Ressmeyer/CORBIS, 36

© NASA/Science Source/Photo Researchers, 27

© National Portrait Gallery, Smithsonian Institution/Art Resource, 16

Brandy Noon, 14, 18, 28, 32, 34

© Reuters NewMedia Inc./CORBIS, 37

© Scala/Art Resource, 7

About the Authors

P.M. Boekhoff is an author of more than a dozen nonfiction books for young readers and she has illustrated many book covers. In addition, Ms. Boekhoff creates theatrical scenic works and other large paintings. In her spare time, she writes poetry and fiction, studies herbal medicine, and tends her garden.

Stuart A. Kallen is the author of more than 150 nonfiction books for children and young adults. He has written extensively about Native Americans and American history. In addition, Mr. Kallen has written award-winning children's videos and television scripts. In his spare time, Mr. Kallen is a singer/songwriter/guitarist in San Diego, California.